THIS WALKER BOOK BELONGS TO:

Jimmy parker
cclelland

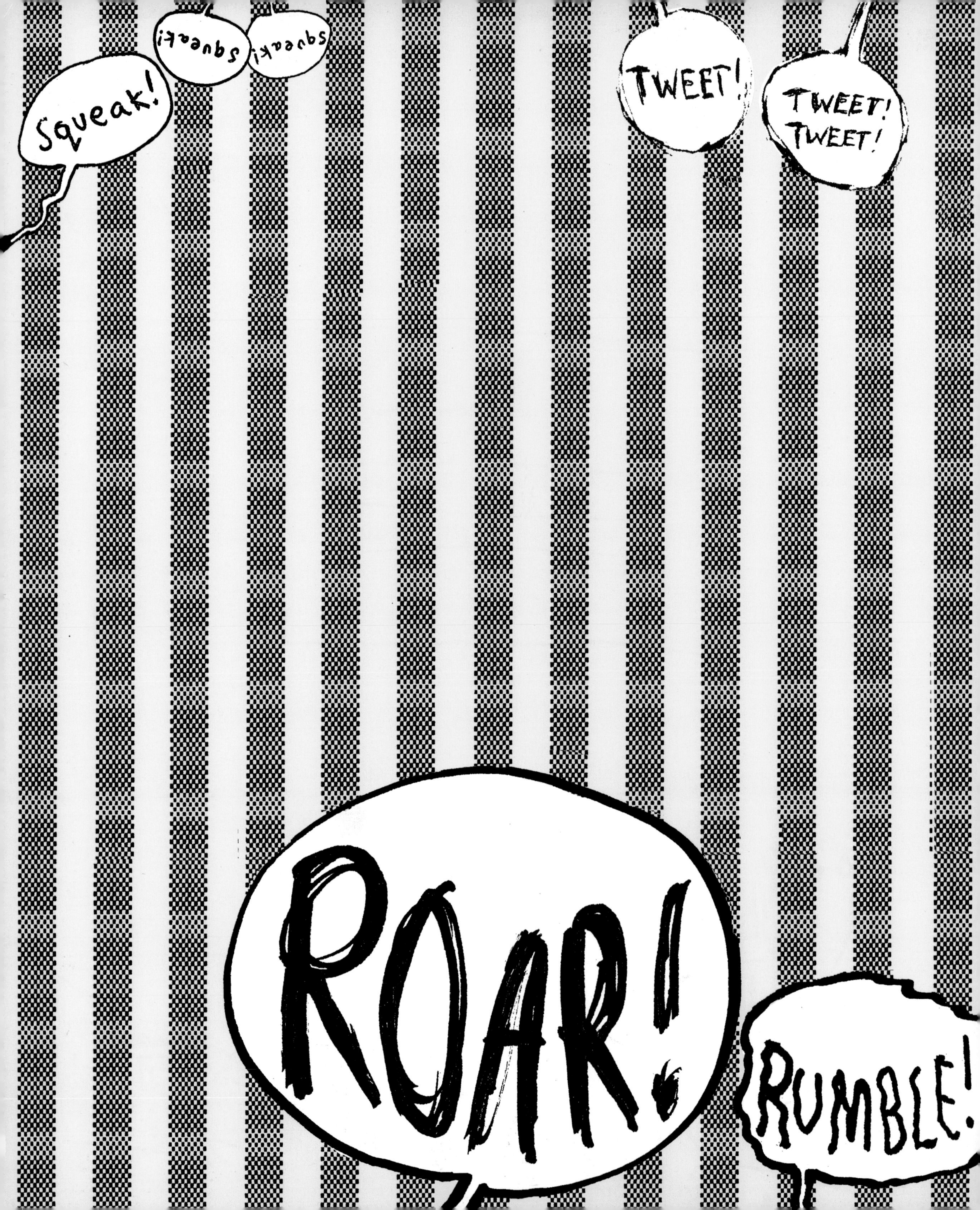
Squeak!
squeak!
squeak!
TWEET!
TWEET! TWEET!
ROAR!
RUMBLE!

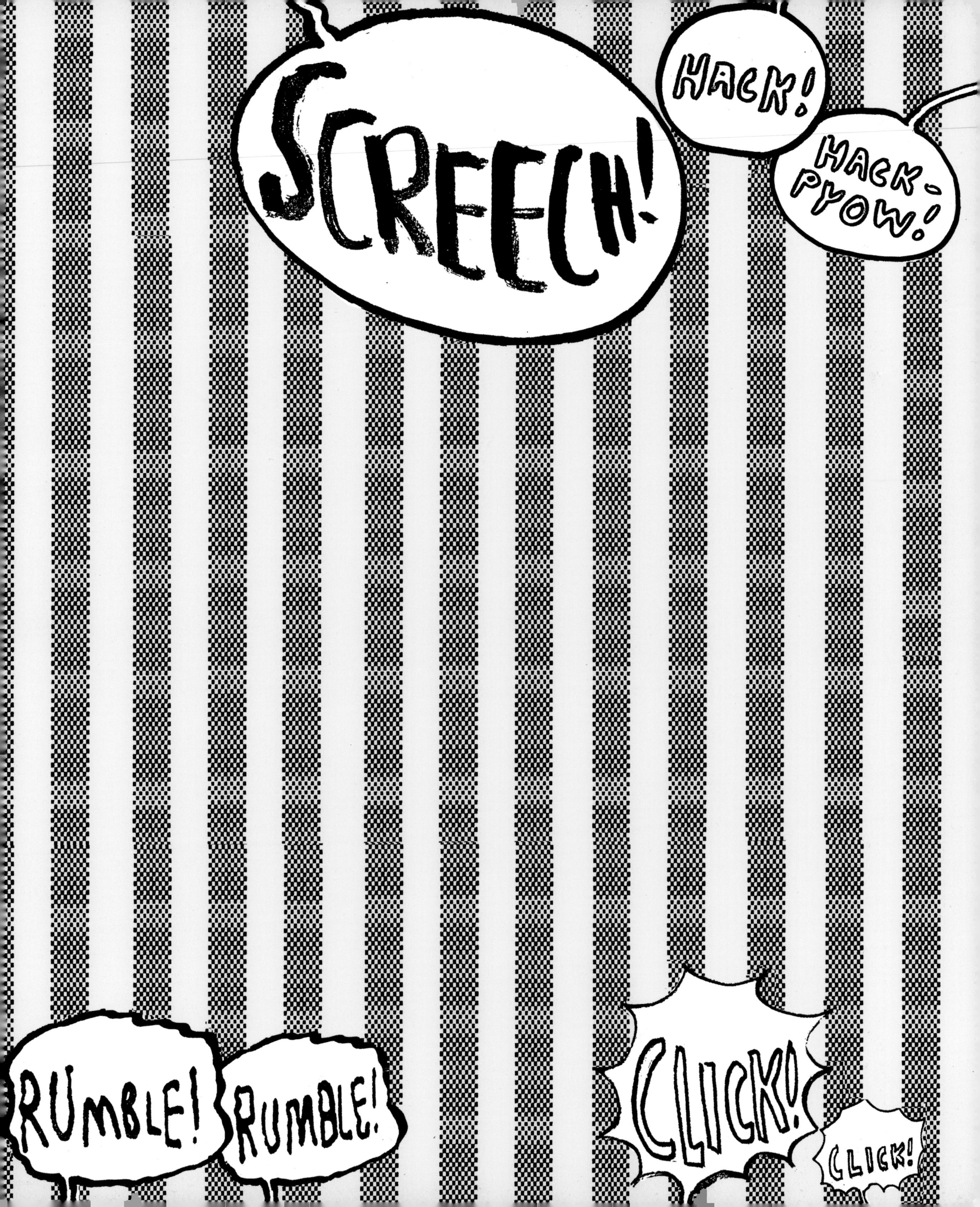
SCREECH!
HACK!
HACK-PYOW!
RUMBLE!
RUMBLE!
CLICK!
CLICK!

THE PANDA on THE OPPOSITE PAGE IS DOING A HANDSTAND AGAINST A TREE you can find out why if you turn to PAGE 14.

For Kieran Maye and other boys who like reading (and those who don't) N. D.

For Ben, the graphic communicator! N. L.

First published 2009 by Walker Books Ltd, 87 Vauxhall Walk, London SE11 5HJ

This edition published 2014

10 9 8 7 6 5 4 3 2 1

Text © 2009 Nicola Davies   Illustrations © 2009 Neal Layton

The right of Nicola Davies and Neal Layton to be identified as author and illustrator respectively of this work has been asserted by them in accordance with the Copyright, Designs and Patents Act 1988

This book has been typeset in AT Arta

Printed in China

British Library Cataloguing in Publication Data: a catalogue record for this book is available from the British Library

ISBN 978-1-4063-5748-6

www.walker.co.uk

Talk, Talk, Squawk!

How and Why Animals Communicate

by **Nicola Davies**

illustrated by **Neal Layton**

WALKER BOOKS

AND SUBSIDIARIES

LONDON • BOSTON • SYDNEY • AUCKLAND

TALK, TALK, TALK...

Human beings never stop communicating!

With words ...

with faces ...

with hands ...

with signs, and signals, flashing lights and sirens.

And we aren't the only ones…

...EVERYBODY'S DOING IT

Wherever you go on the planet, animals are doing it too!

In African jungles, putty-nosed monkeys use sounds just like words: "pyow!" means leopard, "hack" means eagle, and putting the two together, "pyow-hack", means "Let's go!"

Stink bugs in parks and gardens tap out messages on leaves and branches so that they can find one another.

Parasol ants lay scent trails in South American rainforests which act like direction signs, showing other ants the way to the nearest food.

COMMUNICATE OR PERISH!

It's clear from all the singing, tapping, buzzing, humming, flashing and howling going on that communication is as important to animals as it is to us. In fact, almost all the things that animals do – like finding food and shelter, getting a mate and rearing young – would be impossible without it. Every animal has to communicate in some way.

This means they have to send a signal that the animal on the receiving end is going to be able to sense and understand. For instance, it's no good flashing a light to attract a mate if that mate is blind, or whispering in the middle of a thunderstorm, or waving a white flag if the enemy thinks that the sign of surrender is a blue flag. That way both sides lose out.

So animals have come up with ways to make signals that work even when it's too dark to see, too noisy to hear or the animal they're trying to reach is a hundred miles away or even a member of a different species.

In American woodlands, red-eyed vireos sing from dawn to dusk. Each male will sing as many as 22,000 times a day to tell other vireos to keep out of his territory.

Far out in the ocean, blue whales use deep hums, as loud as an aircraft taking off, to send messages across hundreds of kilometres of ocean.

In the murky water of African rivers, elephant-nose fish use electric pulses to send signals that say "I'm the boss".

On coral reefs, the bright colours and patterns of different species of butterfly fish are like labels saying what kind of fish they are.

UNIFORMS

Uniforms are a very simple form of signal that says "one of us". Even the vilest school uniform — like the purple and yellow striped blazer I had to wear to school — is a really useful way to spot your mates in a crowd.

The bright colours and patterns of butterfly fish do the same job as my blazer did, "labelling" butterfly fish so they can find their friends, fast. There can be several different kinds of butterfly fish living on the same bit of coral reef. No fish wants to waste time courting a member of another species, so they all have a different "uniform": colours and patterns that are particular to each species, the way my horrid blazer was particular to my school.

Many other kinds of animals have "uniform" markings that help them to pick out members of their own species. Guenons, a family of monkeys that live in the forests of Africa, have striking facial patterns, so they can find members of their own species when several different kinds of monkey are feeding in the same bit of forest.

Uniforms not only tell you who's on your team, they can tell you what job people do. How would you tell the cops from the robbers if they all wore jeans and T-shirts?

Cleaner wrasse are little reef fish which make their living by, yes, you guessed it, cleaning. They pick tiny parasitic creatures from the skins of much larger reef-living creatures and eat them. This job requires trust on both sides: the wrasse must trust their big customers not to make a meal of them, and their customers must trust the wrasse not to take a mouthful of tender skin. So the wrasse signal that they are not just any old fish out for a quick meal by "wearing a uniform": their bodies are coloured with neat black and blue stripes, making them look quite different from any other fish. To make doubly certain their customers know just who they are, they do a little dance, wiggling their bodies back and forth in the water.

STRIPES SPELL DANGER

There's one kind of animal "uniform" that sends a signal that's understood even by humans. Bright colours or bold stripes spell "danger". Many small animals use this signal to tell the world that they are not a tasty meal!

The bold yellow and black stripes of bees and wasps remind predators that they carry a nasty sting; cinnabar moth caterpillars are packed with poisons from the plants they eat, so to birds their orange and black pattern says "Yuck!"; coral snakes warn of their lethal bite with red, yellow and black stripes; and tiny poison arrow frogs, whose skins carry deadly toxins, are coloured like bright jewels.

Why bother being bright and stripy if you already have a nasty bite, sting or taste to keep predators at bay? Well, firstly, bright colours stand out, so you don't get eaten by mistake because the predator didn't see you or thought you were ordinary tasty prey. Secondly, having a colour and pattern similar to other kinds of inedible or poisonous animals means that the predator may already have learned what your pattern means, without having to sample you to find out.

A "danger" signal like this, that many kinds of animals use and understand, works for everybody: bright, stripy prey doesn't get eaten, and predators don't waste time trying to eat stuff that bites back.

TEAM SMELLS

Colours and patterns that label an animal as "one of us" or "dangerous" are fine for use in daylight, but what happens in the dark?

Poisonous tiger moths come out at night and their main predators are bats, so they use sound to send the "Yuck! Don't eat me!" signal. When a tiger moth hears a bat's echolocation sounds, it makes a high-pitched "click", which a bat can hear. As soon as the bat hears the click, it knows it's chasing something that won't taste very nice, and turns away.

Beavers live in big family groups and don't like sharing their food or shelter with beavers that aren't family members. Inside their "lodge" (the living space inside the beavers' dam), it's too dark to see. So instead of a team colour or pattern, beavers have a team smell. It's a mixture of the smells of all family members and can have as many as fifty different ingredients – so it's too complicated to imitate. Any beaver without the family pong will get chased out of the lodge!

Many mammals have a "team smell" that labels family members as "us" and any animal without it as "them". When your pet cat rubs its head against you, it's swapping smells with you to show you are part of its team.

KEEP OUT

Unlike a sound or a flash of brightly coloured body, smell can go on delivering a message when the animal that sent it isn't there. So smell is the ideal way to say "Keep out". Instead of having to patrol the edges of its territory all the time, an animal can leave a few smelly markers and walk away. They work like a row of "No entry" signs.

This is handy for small animals with big territories to patrol, like desert iguanas. These little reptiles are the size of a skinny pencil case and defend territories as big as half a tennis court. They manage it by squeezing blobs of smell from a row of holes on the underside of each back leg. Each blob is coated with a waxy layer, to stop the smell escaping too quickly. Unlike the desert sand around it, the wax also absorbs ultraviolet light, which iguana eyes are good at seeing, making the scent marks stand out like beacons.

OI! CAN'T YOU SMELL?

THIS IS PRIVATE!

DESERT IGUANA SQUEEZING OUT A BLOB.

KEEP OUT ENOR- MOUS PANDA LIVES HERE!

Pandas have very big, forested, mountain territories. They mark trees on their borders, with scent from a gland under their tail. But they like to send an extra signal: "The panda who left this message is very BIG indeed". So they do handstands, to get their bottoms as high up the tree trunk as possible!

A stale smell sign shows that this bit of territory isn't very well patrolled and could be a good place to invade. So animals never miss a chance to scent-mark their territories, as anyone who has ever walked a dog knows: on a thirty-minute walk a dog may wee in ten different places, leaving a clear "This is mine!" message every time.

NO REALLY ... I SAID "KEEP OUT"!

Sometimes smell just isn't enough to keep intruders at bay and a stronger signal is needed. Wolves live in big extended family groups and back up their territorial scent marks with howling. The whole pack joins in, even the pups, to make the loudest sound possible. Neighbouring packs listen carefully to each other's howls and can tell just how strong another pack is from its howling, so disputes over boundaries can be settled without a fight.

Howler monkeys, too, use howling to keep out of territory trouble. Howlers live in family groups called troops, and if two or more troops run into each other it can end in a quarrel, with family members getting separated and lost. In the tropical forests of South and Central America where howlers live, it's hard to see through the dense greenery and spot where neighbours might be. So first thing every morning the males in each troop howl loudly to show exactly where their troop is. This is such an important part of howler-monkey life that males have a huge throat pouch, just to help them howl extra loudly.

SING IT TO WIN IT

Male birds, too, defend their territories with sound; their songs, which seem so sweet and musical to our ears, are really saying, "Keep out or else!" to rivals and "Come to me!" to potential mates. Every time a territory owner hears another male of his species singing, he must sing back, loud and clear; if he doesn't, his mate may leave him or his territory may get taken over by another male. Singing is so important that some birds sing their songs thousands of times a day.

Even though all bird songs are delivering the same, pretty simple message, songs can be very complicated and every species has a different one. This is partly because of the different habitats birds live in: woodland birds tend to have lots of clear whistles in their songs, as these carry well in the still air of a forest; but grassland or heathland birds have songs with short trills, which carry best in the gusty winds of open habitats. Different species also have different songs for the same reason that butterfly fish are "labelled" in different colours — no one wants to waste time courting or fighting a member of another species, so songs need to be easy to recognize. Finally, although all birds of the same species sing the same song, individual birds sing that song in slightly different ways, so birds can tell their neighbours from new intruders.

SIMPLY DIVINE

Female birds can be very picky about who they mate with, and want only the best fathers for their chicks. So for many male birds "I am gorgeous" is a more important message to get across than "Keep out". All kinds of male birds have come up with wonderful ways to say this to their females, combining flamboyant feathers, fabulous singing and fantastic dance moves, and the most fabulous and fantastic of them all is the male superb lyrebird.

First of all he builds his stage, a flattened circle of earth the size of a large table. Then he starts his song-and-dance routine: he raises his huge tail of lacy feathers like a curtain over his head, shimmying so that it shakes and shivers as he jumps about; he sings, making his song as complicated as possible by copying, perfectly, any sounds he hears — the calls of twenty different kinds of bird plus human sounds like car alarms, chainsaws, even camera shutters — and including them in his song.

And he'll keep the whole performance going for hours on end. It's not surprising that a successful male can attract as many as six different brides to each show, because only a really strong male could be so gorgeous for so long!

17

OTHER WAYS TO SAY "I AM GORGEOUS"

Male blue birds-of-paradise hang upside down while making a noise like a sewing machine to attract females.

Great bustards inflate their throat pouches and stick out their feathers so they look like a large white balloon that can be seen for miles.

Male Anna's hummingbirds do perfect dives and make a buzzing sound with their special tail feathers to look good to the girls.

The female red-necked phalarope mates with lots of males, so she's the one who says "I'm gorgeous" with her bright feathers.

Male humpback whales sing long, complicated underwater songs to say "I'm gorgeous" and a lot more besides which scientists don't yet understand.

Male fruit flies flash their wings in complicated semaphore to signal to their mates.

WHERE ARE YOU?

It's all very well trying to tell your potential partner how wonderful you are but first you have to find each other, and for some animals that can be very tricky.

Elephant-nose fish live at the bottom of rivers in West Africa, where it's far too muddy to find a mate by sight, so they use electricity. They have an electric organ, like a row of batteries, in their tail, which makes patterns and pulses of electricity that are different in males and females. If a male picks up a female's electric buzz with his electric-sensing little "trunk", he'll "buzz" back at her with a typical male pattern.

It's their way of saying "Where are you?" and "I'm over here!".

For a stink bug the size of your little fingernail, finding a mate in an ordinary garden would for a human be like having to search the whole of Texas for a date. Stink-bug males make the search easier by wafting smell signals in the air. Females follow the scent to the nearest plant, but which leaf is her suitor sitting on? He guides her to him by tapping on his leaf, and the taps travel all the way through the stems and leaves of the plant; she picks up the vibrations through her legs, then taps back and, by exchanging signals, they home in on each other!

LONG-DISTANCE CALLS

If you are trying to find a mate, but you don't know where he or she might be, then the further your message can travel, the better. Mole crickets make chirrup sounds to attract a mate by rubbing the edge of their wing on their leg. Like most other cricket chirps, you can't hear them more than a couple of metres away. So mole crickets build burrows and make their calls from there. The shape and size of the burrow mean that it acts just like a trumpet, making the cricket's voice sound 250 times louder — loud enough, in fact, to make the earth round the burrow vibrate and to be heard more than 600 metres away!

Sound travels four times faster in water than in air, so sound is a really good way of sending messages under water, and lots of marine creatures use it. But high, squeaky sounds don't travel as far as low ones. So black drum fish use very low, throbbing sounds to call to one another in the mating season. Unfortunately these fishy love songs can also travel through the walls of seaside houses and keep people awake all night.

The bigger you are, the easier it is to make very low sounds — so blue whales, the biggest creatures on earth, also make the lowest sounds. Their low hums travel brilliantly through water, so they can send a "Here I am" message across a whole ocean.

HAPPY FAMILIES

Those signals "Where are you?", "Come here!" and "I'm gorgeous" are enough to get males and females together to mate. But caring for offspring requires a whole lot more communicating.

Seahorses, unlike most fish, are great parents; but unusually, it's the males who get pregnant. Females put their eggs in the male's brood pouch, where they stay until they are ready to pop out as tiny "seafoals". As soon as one lot of babies leaves Dad's pouch, Mum is ready with a new lot of eggs. To manage this tight teamwork, the couple must communicate well, so they meet up every morning at dawn to dance with their tails entwined.

Most mammal mums do the babycare alone, but in the case of emperor tamarins – tiny mustachioed monkeys from South America – Mum makes sure that she has the help of two dads, because each of her twin babies can have a different father. So when she wants a rest, she signals to the double dads by sticking out her long pink tongue; they know this means it's time for them each to take a twin.

Incubating eggs and feeding chicks is a huge job, so mum and dad birds have to be a good team and it's essential that they keep their pair bond strong. So bird couples spend a lot of time saying "Let's stay together" by greeting, preening and displaying to each other.

TALKING TO MAMA

Babies may start to communicate with their parents even before they are born. Anyone who's kept chickens will tell you how chicks still in the egg start to "peep" to their mum a couple of days before they are ready to hatch. This tells Mum to keep sitting on the eggs as her patience will soon be rewarded!

Baby crocodiles, too, start to call "umph, umph, umph" when they are still inside their leathery-shelled eggs, buried in mud and rotting plants. These piercing little "We're ready!" cries let all the brothers and sisters in the nest know that it's time to hatch, so no one gets left behind. They also trigger help from Mum. When she hears her children's voices she comes to the nest and helps to dig the babies free, then carries them to the water in her mouth.

About the most important message that babies need to give their parents is "Feed me now!". Chicks in a nest have a very simple way of doing this: they open their beaks and start cheeping as soon as a parent bird is near. To make the message extra clear, the inside of a chick's mouth is often brightly coloured. This works brilliantly: the parent birds find a colourful gaping beak irresistible, and they just have to put food in it.

WHERE'S MY BABY?

Feeding babies is hard work, and few animal parents want to waste effort on babies that aren't their own. Babies don't want this either ... if their parents feed a stranger, they might not have enough food for their own children! So where the young of many different parents are mixed up together, parents and their babies are very careful to keep in touch and prevent any mix-ups.

Every night in the breeding season, one and a half million free-tailed bats fly from their roost under Congress Avenue Bridge in Austin, Texas, leaving their hundreds of thousands of babies behind. Each mum bat returns at least once a night to suckle her single baby, and she gives it as much

as a quarter of her body weight in milk every 24 hours. She makes sure that precious milk goes only to her own baby, by recognizing its little squeaky contact call. Mother bats know their baby's voice among all the thousands of others in the same bit of bat roost.

Fluffy grey king penguin chicks at the South Pole must make sure their parents can find them, among thousands of other grey fluffy chicks huddled together for warmth, when they come home with a meal. They, too, do it by voice, and can tell their parents' voices from those of all other adults, even when the howling Antarctic wind drowns out half the calls.

23

GOO GOO, GA GA

Baby birds don't have to learn how to "gape", and parent birds don't have to learn that it means "Feed me now!". A lot of simple animal signalling is like this, with animals born knowing how to send signals and how to understand them. But some more complicated kinds of communication – like singing, for instance – need a little practice.

Chaffinches are handsome little garden birds, whose song is one of the lovely sounds of spring all over Britain and Europe. But chaffinches have to learn their song. Wild chaffinches listen to the songs of lots of other males, and then they practise very quietly, with their beaks closed. At first it's a muddle of too many notes, but after a few weeks of listening and practice, each male chaffinch is ready to sing his song: it's similar enough to that of other chaffinches to say *what* he is, but with enough of his own composition in to say *who* he is.

Dolphin calves click and whistle almost from birth and make a wider range of sounds than adult dolphins do. Scientists think they may be "babbling" like human babies. From the age of about seven months a human baby tries out all the different sounds he or she can possibly make, to see what happens! Useful sounds – like "Mama" or "Dada" – are remembered, but sounds a baby doesn't hear spoken (because they belong in a language Mum and Dad don't speak) or that don't have an effect get forgotten. After a few months baby dolphins too have dropped some sounds and kept others, so both human and dolphin babies end up with a tool kit of sounds that match the grown-ups around them.

LIVING TOGETHER

Good communication helps mates find each other, helps parents bring up babies and keeps communities working together.

Honey bees live together in colonies of up to a hundred thousand. There is usually just one Queen who does nothing but lay eggs. All the other bees are her children, mostly (female) workers, which tend the Queen, look after the eggs and larvae, clean out the hive and fly off in search of nectar and pollen to keep everybody well fed. Everything is perfectly ordered, because of a very clever system of chemical communication. The Queen and the workers swap cocktails of chemical messages called pheromones, which control the behaviour of all the bees in the colony, making sure everyone does the right job at the right time.

The bees also use dance to communicate with one another. A foraging bee that has found nectar-loaded flowers needs to tell as many other workers about it as possible so they can gather lots of nectar for the colony. So she dances, waggling her bottom as she crosses the comb inside the hive. Other bees touch her with their antennae, feeling the direction of her dance, the number of her waggles and how excited she is. These things tell them where the nectar can be found. Then they make a "beeline" for the flowers.

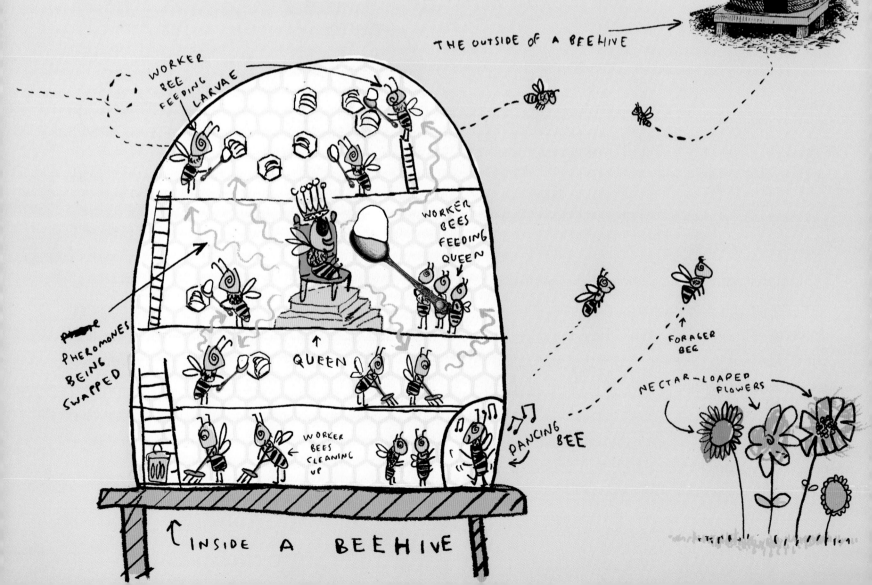

THE OUTSIDE OF A BEEHIVE

WORKER BEE FEEDING LARVAE

PHEROMONES BEING SWAPPED

WORKER BEES FEEDING QUEEN

QUEEN

WORKER BEES CLEANING UP

DANCING BEE

FORAGER BEE

NECTAR-LOADED FLOWERS

INSIDE A BEEHIVE

JOBS LIST 9.00 BUZZ 9.30 FORAGE 10.15 BUZZ 12 PM FEED

RUMBLES IN THE JUNGLE

African elephants also live in family groups led by a female, the matriarch, and their survival depends on their working together. Food and water can be scarce and predators can threaten calves. The matriarch is usually the oldest elephant in the group. Making use of her long memory about where to find food and water when times are hard, and the ability of the herd to stick together, are very important, and both rely on good communication.

THE MATRIARCH →

Researchers have found that elephants use 160 touch and visual signals, and another 70 sound signals. Some of these sound signals work in an extraordinary way. Elephants make rumbling sounds well below the range of human hearing. Like all low sounds these are very good at travelling quite far in the air, but they also travel like tiny earthquakes through the ground, and elephants feel them with their sensitive feet and trunks. In quiet areas of Africa, elephant rumbles can call the herd together over several miles. This could certainly explain how, when part of the herd is in trouble, the rest of the gang appear out of nowhere as if by magic.

27

COMMUNICATION: THE DARK SIDE

Wherever there is communication there is also lying, and I'm sorry to have to tell you that there are some liars in the animal world. Hoverflies have the bright black and yellow stripy pattern that says "danger", just like a wasp. But hoverflies are lying: they have no sting. Milk snakes have a warning pattern of red, yellow and black stripes like the venomous coral snake, but milk snakes are harmless. Milk snakes and hoverflies are using their lies in self-defence – but there are much more lethal liars.

The fangblenny can make itself look and behave just like a cleaner wrasse, advertising cleaning services to reef fish. But its customers soon find out the fangblenny is lying, when it bites a chunk out of them. The female bolus spider wafts a message-carrying chemical, called a pheromone, in the air that is exactly like the one female moths use to say – "Here I am come and get me". Male moths fly up eagerly, expecting a hot date, and instead end up as a spider's supper.

But there aren't many liars, because if this kind of lying got too common everyone would lose out. The meaning of the original message would change: a hoverfly's stripes would say "dinner" not "danger", and the fangblenny's uniform would not say "Trust me" but "Eat me before I eat you". Everyone, both liars and truth-tellers, could get eaten.

TRUTH AND LIES

Believing a lie can get you into trouble, as we've seen from the bolus spider. A female who believes a male who lies about how big and strong he is can end up mating with a wimp; a male who believes the same lie might give up and lose a fight that he could have won. So animals are on the look-out for lies, and for honest signals that can't be faked.

Every autumn, red-deer males compete to mate with the most females. Fighting with those big pointy antlers is jolly dangerous, though, so they try to decide which of them is strongest by showing off. They roar at one another, loud, deep, long roars that say "I'm big and fit and strong". The great thing about roaring is it can't be faked: it's

hard work, and to make a big, strong sound you have to be a big, strong deer. Many contests don't get further than roaring, as weaker males can hear how strong the opposition is and know it's not worth risking injury in a fight. Only males that are so evenly matched that their roars sound the same settle their rivalry with a real fight.

MONKEY BUSINESS

So far we've talked about messages and signals, which, like road signs, are fine for simple communication. But you couldn't have a proper conversation using traffic signals. For really complicated communication about ideas and feelings, places and things, you need language, and one of the first things a language needs is words. Words can be made of sounds, gestures, smells, visual displays or anything that an animal's senses can detect. For example:

The sounds "k", "a" and "t" together mean CAT, and  means the same in sign language.

But CAT and  are always the same and always mean the same thing.

Humans are certainly not the only ones to have words. Vervet monkeys have different alarm calls for different sorts of predators, such as eagle, snake and leopard. When vervets hear the "snake" call, they all stand up on their hind legs and look down at the ground, carefully watching for a slithering python. When they hear "leopard" they leap high into the trees where a leopard couldn't catch them, and when they hear "eagle" they dive into dense bushes where an eagle couldn't spot them.

Several species of African monkey have been found to have predator "words", and different species living in the same area understand each other's predator calls. This is a big advantage as it means lots more pairs of eyes looking out for danger, and lots more voices raising the alarm.

Words alone aren't enough; to make a language they need to be strung into sentences, with a more complicated meaning, and it looks as if monkeys do this too. They add other calls to the predator alarm "words", which, judging by the reaction of the troop, say things like "right now!" or "far away".

To try and find out what monkeys are saying, researchers record their calls, then play them back and see what the troop does. This is tricky and time-consuming, and doesn't work very well for the quieter calls that monkeys use for one-to-one communication. But if monkeys can say "Eagle up above right now!" and "Leopard quite far away. Not urgent", who knows what else they might talk about? They can certainly use words to lie. A monkey that doesn't want to share a particularly yummy food item might call out "Snake, right now!" to send the rest of the troop running for the treetops, leaving the liar to eat its special food-find in peace.

TALKING ANIMAL

Scientists, and people who study language and communication, argue about whether animals, like those monkeys, have a language in the way we do. If they do, then perhaps they could learn a bit of ours. Some researchers think that there are animals that have already learnt "human".

Chimps are our closest relatives, but they don't have vocal cords, tongues and mouths like us, so they could never form words. But their hands and many of the gestures they use are very like ours. So in the 1960s an experiment was begun to teach a young chimp, called Washoe by her human carers, American Sign Language, the language used by Americans with hearing or speech difficulties. Washoe learned 250 different signs and taught another chimp 50 signs, without any human help at all. Washoe could make sentences, asking for food, play or drink, and would put signs together to help name or describe things she'd never seen before. Her little group of captive chimps used normal chimp sounds, gestures and touch to communicate with one another, but also used human sign language.

Parrots are wonderful mimics so it's not difficult for them to say human words, but Alex, an African grey parrot that lived with researcher

Irene Pepperberg for 30 years, seemed to understand the 150 words he learnt to say. Alex could answer questions about the shape, colour, size and number of objects placed before him, even when he'd never seen them before. Like Washoe, Alex sometimes made sentences on his own that seemed to show that he really understood the words he was using; when Irene once had to leave him at the vet for treatment, he said, "Come here. I love you. I'm sorry. I want to go back."

Some people say that Washoe and Alex didn't really understand what they were saying and that their words tell us nothing about animal thoughts and feelings. But Washoe and Alex were both far from their wild homes, living with humans, not their own kind. That's like you or me learning "alien" in a flying saucer on the way to Mars, and I wonder how well we'd manage? Our human language has evolved over tens of thousands of years to express what goes on in our world, in our heads and our hearts. But animals are not people; they have their own thoughts and feelings very different from ours. How could our words ever fully express what they might want to say?

Scientists listening out for alien communications from space have developed computer programs that can spot the tell-tale footprint of a language in any signal. One day this technique could be used on dolphin sounds, which could be close to our idea of language. Maybe then we'll finally know what animals have to say about us.

INDEX

GLOSSARY

Colony a group of animals that live together and share the jobs they need to do to stay alive.

Echolocation sounds are made by bats and some other animals. They create echoes that the animal can use to find its way around.

Evolution how animals and plants change over many generations, sometimes enough to become whole new species.

Habitat the kind of place where you would find a particular animal. For example, a whale's habitat is the ocean, not the forest.

Incubating birds sit on their eggs to keep them warm so the chicks inside can grow. This is called incubating. Reptiles incubate their eggs too, by burying them somewhere warm, or keeping them inside their bodies.

Matriarch the female leader of a group of animals, usually older and wiser than the others; a kind of super granny!

Pheromones a signal-carrying chemical produced by an animal, that switches on a particular bit of behaviour when sensed by other members of its species.

Semaphore type of signalling using flags held in different positions to convey letters and words. Fruit flies hold their wings in different positions to send messages too.

Species a kind of animal or plant. A dandelion is a species of plant and a zebra is a species of animal.

Territory the area an animal uses to find food, shelter and mates and doesn't want to share with others!

Toxins another word for poisons.

Ultraviolet a kind of light, just beyond blue in the rainbow that humans can't see very well but insects and other animals can. It's the part of sunlight that makes your skin go brown.

ABOUT THE AUTHOR

Nicola Davies is an award-winning author, whose many books for children include *The Promise*, *A First Book of Nature*, *Big Blue Whale*, *Dolphin Baby*, and *The Lion Who Stole My Arm*. She graduated in zoology, studied whales and bats and then worked for the BBC Natural History Unit. Visit Nicola at **www.nicola-davies.com**

"When I was little I wished I could talk to animals," Nicola says, "but writing about how they talk to each other is the next best thing!"

ABOUT THE ILLUSTRATOR

Neal Layton is an award-winning artist who has illustrated more than sixty books for children, including the other titles in the Animal Science series. He also writes and illustrates his own books, such as *The Story of Everything* and The Mammoth Academy series. Visit Neal at **www.neallayton.co.uk**

About this title, he says, "Since illustrating this book I'll never look at any twittering, tapping, rumbling, roaring, buzzing or brightly-coloured animal in the same way again!"

SOURCES

If you'd like to read more about Alex the Parrot and how he learned to talk:

Alex and Me by Irene Pepperberg (Harper Perennial, 2009)

This is a grown-up book but written by a scientist who is really good at explaining things to non-scientists:

Animal Communication by Phil Gates (Cambridge University Press, 1997)

This is a book about bird behaviour and intelligence but also about how birds communicate, written by a scientist who has spent a lifetime studying them and telling people about them:

Bird Sense by Tim Birkhead (Bloomsbury Paperbacks, 2013)

I'm coming dears...
mama!
mama!
KEEP OUT!
SAME TO YOU!
I AM A HUNGRY LION!
QUICK, CALL THE HERDS...!